ADIRONDACK ALPINE SUMMITS

An Ecological Field Guide

by Nancy G. Slack & Allison W. Bell

Front cover photographs: alpine summits of Mounts Colden, Marcy, and Skylight from treeline on Algonquin Peak; Lapland rosebay, diapensia, alpine azalea, harebells by Allison W. Bell. Magnolia warbler by Jeff Nadler, short-tailed weasel by Olivier Gilg, black-backed woodpecker by Steve Faccio. Back cover photographs: snowshoe hare by Eric Dresser, yellow-rumped warbler by Jeff Nadler, bog laurel and map lichen by Allison W. Bell.

Book design by Allison W. Bell. All photographs © Allison W. Bell unless otherwise noted.

Published by the Adirondack Mountain Club, Inc.
814 Goggins Road, Lake George, NY 12845-4117 · www.adk.org

Our thanks to Eric Dresser, Steve Faccio, Jan-Peter Frahm, Olivier Gilg, Warren Greene, Patrick LaFreniere, Jeff Nadler, Bill Silliker, Peter Zika, and the Adirondack Museum for providing images for this book.

Library of Congress Cataloging-in-Publication Data

Slack, Nancy G.
Adirondack alpine summits : an ecological field guide / by Nancy G.
Slack & Allison W. Bell. — 2nd ed.
p. cm.
Rev. ed. of: 85 acres. c1993.
Includes bibliographical references and index.
ISBN-13: 978-1-931951-18-0 (pbk.)
ISBN-10: 1-931951-18-7 (pbk.)
1. Natural history—New York (State)—Adirondack Mountains. 2. Mountain ecology—New York (State)—Adirondack Mountains. 3. Adirondack Park (N.Y.)—Guidebooks.
I. Bell, Allison W. (Allison Williams), 1957- II. Slack, Nancy G. 85 acres. III. Title.

QH105.N7S59 2006
508.747'5—dc22

CONTENTS

This book is dedicated to
the late Barbara McMartin and to Ruth Schottman.

Barbara McMartin inspired us with all of her books,
especially *The Great Forest of the Adirondacks,* and her innovative
trail guides. She was a thoughtful friend and a constant
advocate for Adirondack preservation.

Ruth Schottman has introduced a great many hikers to the
wildflowers and natural history of the Adirondack High Peaks with
her ADK workshops, her articles, and her book, *Trailside Notes:
A Naturalist's Companion to Adirondack Plants.*
She has been a wonderful field companion.

The Adirondack high country—what a grand expression! These four words summarize the spectacular vistas and exciting environment to be experienced on the mountains that crown the geography of the Empire State. The several highest Adirondack summits are open grandstands for thrilling scenery, as well as natural museums in which hikers may observe the specialized vegetation and post-glacial conditions that still exist on these islands in the sky.

"Ketch" Ketchledge with alpine ecology students atop Whiteface Mountain, summer 1993.

The fragile life zone on these mountaintops is but two feet thick, the distance between the impenetrable bedrock and the killing winds tearing over the ground above. This is no place for unknowing visitors to lower the survival odds by trampling rare plants and delicate habitat. Instead, it is a place for wonder and learning, a place to reach inner heights of understanding equal to the grandeur of these ancient peaks.

The Adirondack Mountain Club is performing an important service in publishing this guide to help visitors appreciate the endangered alpine habitat they witness on our precious Adirondack summits.

Edwin H. Ketchledge
Professor Emeritus, SUNY College of
Environmental Science & Forestry

TRAIL TO →
← MILES
FOLLOW TRAIL MARKER MARKERS
AVALANCHE LAKE .7
ADIRONDACK LOJ 5.8

TRAIL TO →
← MILES
FOLLOW TRAIL MARKER MARKERS
MT. ALGONQUIN (elev. 5114' ascent 2300') 1.9
ADIRONDACK LOJ via ALGONQUIN 5.7

LAKE COLDEN INTERIOR HUT
CROSS BRIDGE
TAKE BLUE TRAIL

Onwards and upwards! A climb to an Adirondack alpine summit is a journey through fascinating forest zones.

INTRODUCTION

This book is designed as an introductory guide to the high country of the Adirondacks, especially the unique 85 acres of the alpine zone. It is intended as a nature tour in print and photos—from the mountain trailheads to the summits of the High Peaks. In a region noted for its natural diversity, a climb to any of the Adirondacks' highest summits is a journey through a series of distinct and fascinating habitats. It is our hope that with this guide you will learn something of this area's ecology and its plants and animals, especially those of the summits. We hope in so doing, you will help protect this fragile legacy.

Whichever Adirondack peak you climb, the rock you walk on is more than a billion years old, an ancient dome of uplifted metamorphic rock. It was once covered with layers of sedimentary rock literally miles thick that gradually eroded.

One million years ago, the whole region, even the mountain summits, was covered by glacial ice. When the last of the glaciers melted about 13,000 years ago, no plants or soil remained. The scoured landscape was left with deposits of glacial "till," sand and rock debris, and occasional huge boulder "erratics" you can still

High Peaks bedrock is more than a billion years old.

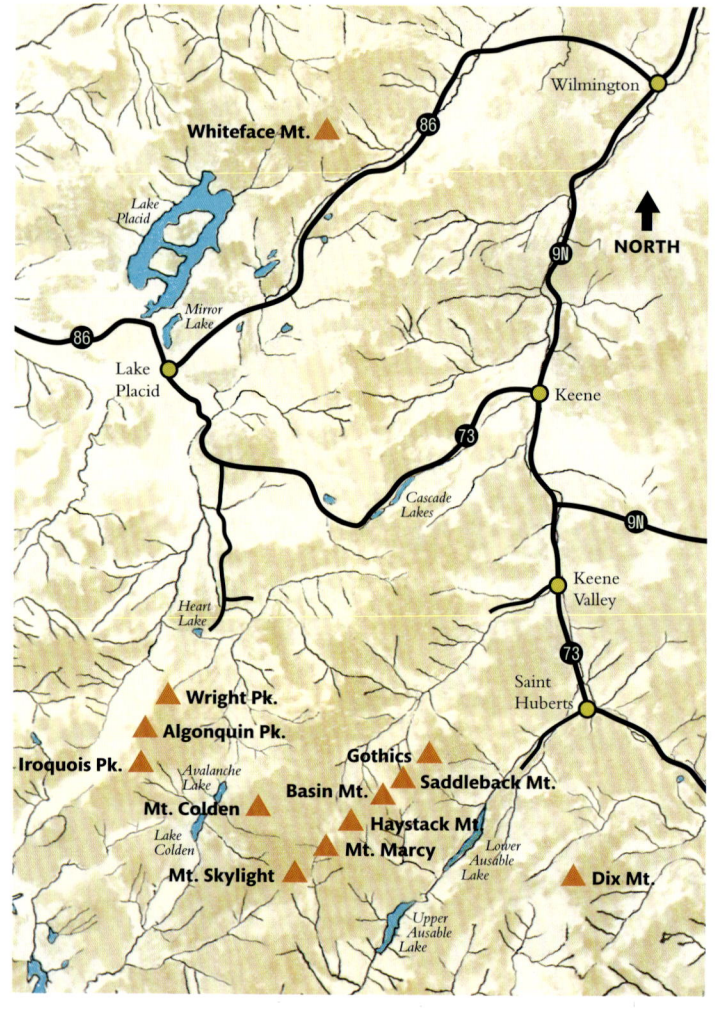

Adirondack peaks with significant areas of alpine habitat

see along the trails. Many regional land forms are glacial legacies—the amphitheater or "cirque" on Giant Mountain and the Cascade Lakes, for example.

As the glacier retreated, an arctic tundra landscape was created. New soils were formed and retained with the help of pioneering lichens and mosses, which were then joined by colonizing arctic plants. Even muskoxen roamed here. As the climate warmed, conifers and eventually sugar maples and other deciduous trees migrated north and colonized the slopes. Spruces and firs dominated the higher slopes, crowding out the once-widespread arctic tundra plants, now left on

Adirondack forests have been evolving for thousands of years and are changing still.

very special "islands" on our coldest Adirondack summits.

Downslope, forests grew and regrew after fires, windstorms, and lumbering. Virtually none of this area is virgin forest; the "old growth" forests are no more than 200 years old. The alpine landscape on the summits, however, has been there for 10,000 years. It is a fascinating living museum.

As you begin your mountain hike, you pass through a character-istic Adirondack forest type, with sugar maples and other deciduous trees. As you climb, most of these

9

broad-leaved trees drop out. At about 2500 feet, you will find yourself in a spruce-fir forest with different flowering plants underfoot and different bird songs up above. At some point above 4000 feet you will enter a zone of gnarled dwarf trees, or "krummholz," which gradually gives way to the alpine zone of the summit, closely related to the arctic tundra many miles northward. Erect trees cannot grow here, but mosses, flowering cushion plants, dwarf willow, and small shrubs form a fragile polychrome carpet.

Of this beautiful and unique habitat only 85 acres exist in the Adirondacks. Along with its

Exposed anorthosite contains opalescent blue labradorite crystals.

spectacular views, this zone offers its own colorful flowers, birds, lichens, and butterflies. With this book in your pocket they are yours to discover and enjoy. They are worth the climb and worthy of your care.

CLIMBING THE HIGH PEAKS

Two European explorers first glimpsed the high peaks in the same year: 1608. Samuel de Champlain explored south down the lake later named for him and saw mountains to the east and south. Henry Hudson sailed his ship *Half Moon* up the river named for him and saw the Adirondack high peaks later that same year. Darby Field with two guides made the first recorded ascent of Mount Washington in 1642, but the first Adirondack peak was not officially climbed for nearly 200 more years.

In 1836, New York Governor William L. Marcy authorized a survey of the yet-unexplored Adirondack wilderness, with geologists Ebenezer Emmons and James Hall in charge. An unofficial survey party got underway first, led by William Redfield. Other members of the party included David Colden, three partners from the McIntyre Iron Works, and woodsman John Cheney, who was to become Mt.

Marcy's first famous guide. Their route took them to Lake Colden, Avalanche Lake, and within sight of Mount Colden, Mount Marcy, and Algonquin Peak. Until then it was thought that the Catskills were New York's highest peaks. But Redfield maintained that Marcy was higher, and vowed to climb it the following year.

In September of 1836, Emmons and Hall climbed Whiteface and measured its height at over 4800 feet! The Adirondacks were higher than the Catskills. Emmons, Hall, and Redfield returned the next summer to officially pursue the geological survey and to climb Marcy. A large party started out, including botanist John Torrey, artist Charles Ingham, and two woodsmen guides,

Cheney, and Harvey Holt. They camped at 3700 feet, and early on August 5, 1837, this group scrambled through the dwarf trees of the krummholz zone and at last reached Marcy's summit, "covered only with mosses and small alpine plants." The summit measured over 5300 feet, more than 1000 feet higher than any Catskill peak. Three days later they reached the top of the McIntyre Range, Algonquin Peak.

In 1838 a 15-year-old girl named Esther McComb had been the first to climb the 4270-foot peak now named for her—Esther Moun-

INTRODUCTION

tain. She and her siblings, who lived near Wilmington, were forbidden to climb the peaks, but she set off anyway for Whiteface—before there was a trail. Instead of Whiteface, she found herself on top of its lower sister mountain. There is a 1939 plaque on Esther Mountain's summit, proclaiming Esther McComb the Adirondacks' first recreational climber.

The climbing scene soon moved to Keene Valley and the celebrated and eccentric guide known as Old Mountain Phelps. He was the first to climb Haystack in 1849, and in 1861 he cut the first trail up Marcy from the Ausable Lakes and through Panther Gorge. He was the

first to climb Giant in 1854, and in 1873 with New York State Surveyor Verplanck Colvin did the first ascents of Skylight and Mt. Colvin.

It was Verplanck Colvin who conducted the most important exploration of the High Peaks. He was Superintendent of the Adirondack Topographical Survey for 28 years. He worked his men hard, too hard in bitter weather, but he was always there with them. His scientific reports were well written and widely read. Colvin and his assistant, Joseph Blake, were fanatic mountain climbers. A great many peaks were summited and measured, including the first ascents of Skylight, Gray Peak, and Upper Wolf Jaw. In 1872, together with guide Bill Nye, Colvin

Climbing Mount Marcy in 1859— no maps, no trails, no fancy field guides.

discovered Lake Tear-of-the-Clouds, the source of the Hudson River. Colvin was also an extremely important advocate for the preservation of the Adirondack wilderness. In 1885 the state legislature created the Adirondack Forest Preserve, which was to be kept forever wild.

In 1880 Henry van Hoevenberg ("Mr. Van") built the first Adirondack Lodge. He bought the land around Heart Lake and cut trails, including those to Marcy and Algonquin, the two highest peaks. The Lake Placid Club, with its own lodge on Mirror Lake, bought van Hoevenberg's lodge in 1900. By then there were also four trails up Whiteface. In 1887 a group of wealthy preservationists from Philadelphia bought the Adirondack Mountain Reserve (AMR), a huge tract of land including both Ausable Lakes and the surrounding mountains, and built the predecessor of the present Ausable Club at St. Huberts. The mountaintops are now owned by the state; trails through AMR land are open to all hikers and climbers. In the 1890s the Adirondack Trail Improvement Society (ATIS) was formed, after a Keene Valley climbing party struggled through downed timber to the summit of Noonmark. ATIS is still active, with a variety of trail programs.

The Adirondack Mountain Club (ADK) was founded in New York City in 1922 with many prominent New Yorkers, including the young Franklin Delano Roosevelt, as charter members. Its stated objectives were to encourage hiking and mountain climbing, and to develop, extend, and maintain public trails and campsites. Guidebooks, maps, excursions, and conservation were also early objectives. Construction of the Northville–Placid Trail (originally the Long Trail) was begun in 1922. The club's other early big project was to construct Johns Brook Lodge in the wilderness above Keene Valley; it opened in 1925. Trails were also planned to Big Slide and the Great Range. With the help of the New York State Conservation Department and other groups many more trails were built in the High Peaks region.

In the 1920s and 1930s camping and backpacking became a major part of climbing, and many women participated. Early climber Grace Hudowalski wrote that she climbed Mount Marcy in 1922 in her "voluminous bloomers"—surely an improvement over botanist Elizabeth Britton's long skirt and high button boots worn for a climb up Whiteface in the 1890s. In the 1930s, Orra Phelps, who wrote some of the first High Peaks trail guides for ADK, described carrying vegetables in her bedroll and eggs in a cooking pail— all before portable camp stoves,

sleeping bags, and lightweight tents became available. Bob and George Marshall and guide Herbert Clark were the first to climb all 46 peaks over 4000 feet; the later (1948) Adirondack 46er organization encouraged many to climb these peaks. It is now a challenge to manage the often overcrowded conditions of trails and campsites, especially in the fall color season. Some of us prefer to climb in June, when the alpine flowers are in bloom and the human population is low—though the blackfly population is sometimes high.

In addition to the early geologists, other scientists have added to our knowledge of Adirondack natural history. Beginning in the 1860s, Charles Horton Peck found many new plants, explored the High Peaks, and became the prime discoverer of mushrooms new to science. Twentieth-century botanists, zoologists, and ecologists including Stanley Smith, Ed Ketchledge, Orra Phelps, Norton Miller, Steve Young, Alvin Breisch, Glenn Johnson, and Nancy Slack have explored and done research in the High Peaks. Animals of the high mountain zones, including mammals, amphibians, and birds, have been monitored. A study of the endangered Bicknell's thrush, which breeds on mountain summits, is underway. Pollination of alpine flowers is another current research topic. It is surprising how many butterflies, bees, flies, and other insects make their way up to 5000 feet! Adirondack summits are not just for us humans!

NORTHERN HARDWOOD FOREST

When you start up the trail at about 1500 feet at Keene Valley or St. Huberts, or just over 2000 feet at Heart Lake, you are in northern hardwood forest. Sugar maple is the Adirondacks' great glory, but beech, yellow birch, and hemlock are also characteristic of the mature forest. In many places the forest has been disturbed and is in an earlier stage of succession. You may find aspens and paper birches predominating here, and a great diversity of plants in the sunlit understory at the trailhead and on lower slopes.

Notice the green-and-white-patterned bark of striped maple; its leaves turn bright yellow in the fall. Hobblebush is a common shrub in this and higher elevation forests. It has large clusters of white flowers in spring and bright red berries in late summer and fall. Canada mayflower, wild sarsaparilla, and the double-decker Indian cucumber-root are characteristic of this zone's ground flora. In spring and early summer you can find blue, white,

Starting out—many summit trails begin in the diverse northern hardwood forest.

and even yellow violets and red and painted trillium along the trail. Pink ladyslipper orchids bloom here in June.

A great variety of birds and other animals live in this zone as well. Walk quietly, especially at dawn or dusk, to see them. White-tailed deer, porcupines, raccoons, and chipmunks are common, but you might also come upon a black bear, a short-tailed weasel, or the largest Adirondack member of the weasel family, the fisher. Watch for amphibians: red efts, "masked" wood frogs, and big American toads. Listen for the "Here I am, where are you?" song of the red-eyed vireo and the varied flute-like song of the hermit thrush. This is a temperate ecosystem of great diversity. If you are camping here, watch and enjoy its other inhabitants.

American beech
Fagus grandifolia
Common to 2500 feet. Toothed, parallel-veined leaves, edible nuts. Formerly dominant. Bark disease mars its grey bark, kills mature trees, but sprouts persist. *Beech family*

Yellow birch
Betula alleghaniensis
Dominant in northern hardwood and transition forests to 3000 feet. Peeling brass-colored bark is distinctive; leaves are similar to paper birch. *Birch family*

Sugar maple
Acer saccharum
Dominant northern hardwood tree; common to 2500 feet. Five-lobed leaves, U-shaped notches between lobes, orange-red in fall. Shade tolerant, prefers rich soil. *Maple family*

Mountain maple   ✿ JUN
Acer spicatum
Small tree; grows with yellow birch in transition zone to 3000 feet. 3-lobed veiny leaves with rounded teeth; upright "candles" of greenish flowers. *Maple family*

Red maple
Acer rubrum
Grows to 3000 feet. Three-lobed leaves with V-shaped notches between lobes, deep red in fall. More tolerant of poorly drained, shallow soils. *Maple family*

Striped Maple   ✿ MAY-JUN
Acer pensylvanicum
Also called "goosefoot" from the leaf shape, or "moosewood" because moose eat it. Small striped-barked understory tree; hanging yellow-green flower clusters. *Maple family*

Eastern hemlock
Tsuga canadensis
Dominant, highly shade-tolerant conifer to 2500 feet. Shaggy appearance. Small single needles in flat sprays, 1" cones. Seeds provide food for wildlife. *Pine family*

Wild Sarsaparilla 🌼 MAY–JUN
Aralia nudicaulis
Grows to timberline. Large basal leaf with 3 groups of 5 leaflets. Separate flower stalk has 3 umbels of green flowers, then dark purple berries. *Ginseng family*

Hobblebush ❀ MAY
Viburnum lantanoides
Common shrub to 3000 feet with opposite, heart-shaped leaves. Large outer sterile flowers attract pollinators; bright red berries in fall. *Honeysuckle family*

Pink ladyslipper 🌸 JUN
Cypripedium acaule
A pouched orchid, found to nearly 4000 feet, in beautiful trailside clumps. Look for the rarer white form. One of five New York ladyslippers. *Orchid family*

Common blue violet ✿ APR–JUN
Viola sororia
Grows in northern hardwood forest, commonly in moist places along trails, often with a small white violet. No leaves on stem. One of seven Adirondack violets. *Violet family*

Round-leaved yellow violet
Viola rotundifolia ✿ APR–MAY
Grows in northern hardwood and transition forests. Heart-shaped leaves with scalloped edges are small at flowering and may grow to 5" in summer! *Violet family*

Trout lily ✿ APR–MAY
Erythronium americanum
Northern hardwood forest to 2500 feet. One of our earliest spring bulb flowers. Yellow bell-shaped flowers and mottled (trout-like) leaves. Also called "dogtooth violet." *Lily family*

Red trillium ✿ MAY–JUN
Trillium erectum
Mainly northern hardwood forest to 3500 feet; 3 maroon petals, 3 green sepals, 3 whorled leaves. Look for a yellow form. Its bad smell attracts carrion fly pollinators. *Lily family*

Indian cucumber-root   ✿ JUN
Medeola virginiana
Found to 2500 feet. Called "double-decker plant." Two levels of whorled leaves are streaked with magenta in fall. Flowers hang down; dark purple berries. *Lily family*

Rock fern
Polypodium virginianum
In northern hardwood and transition forests to 2500 feet. Forms mats on tall boulders where deer can't browse. Evergreen leaves, brown spore cases, no separate leaflets.

Lung lichen
Lobaria pulmonaria
Primarily on sugar maples in old-growth northern hardwood forests (as on Ausable Club land). Very susceptible to air pollution. Turns from brown to bright green in rain.

Shining clubmoss
Hyperzia lucidula
A common clubmoss in shady northern hardwood and spruce-fir forests. Has no spore-cones; look for the yellow spore cases on the shiny green leaves.

White admiral
Limenitis arthemis
Also called banded purple; this butterfly has a striking white wing band. Emerges in June. Can be seen on forest trails on your way to the alpine flowers.

Black bear
Ursus americanus
Only Adirondack bear, usually not dangerous unless one tries to feed or confront it. Can be a nuisance at campsites by stealing food. Omnivorous, eats plants and small animals.

American toad
Bufo americanus
Common trailside insect-eater, also found quite high up. Females often more gaily colored. This species has only 1 or 2 warts in each dark spot; rarer Fowler's toad has 3 or more.

Red eft
Nothophalamus viridescens
The 2- to 3-year land stage of the red-spotted salamander, one of many Adirondack salamanders, often seen after rain. Tadpoles and olive-green adults live in ponds and wetlands.

Starting at about 2000 feet, you see "Christmas tree"-shaped spruce and balsam firs appearing among the deciduous hardwoods. This is the transition forest—between northern hardwoods and spruce-fir. By the time you have hiked up to 2500 feet, many environmental factors have changed and you have entered a very different forest: the coniferous spruce-fir forest, or boreal (northern) zone. Here, the growing season is shorter, temperatures are colder and effective precipitation is higher. Soils are shallower, wetter, more acidic, more organic, and less fertile. Boreal forest species are well-suited to these conditions. The needles of evergreen trees resist desiccation and conserve scarce nutrients. Spruce, fir, and paper birch can withstand winter temperatures below –30° F without injury. This kind of forest is more typical of Canada, but you have reached it in a short climb instead of a long drive north.

In the transition zone between forest types you still find yellow

In the thick of it—the Algonquin Trail through spruce-fir forest at 3500 feet.

SPRUCE-FIR FOREST

Steep mountain streams cut through high-elevation forests.

birch and striped and mountain maple, but evergreen red spruce and balsam fir quickly dominate the boreal forest as you climb higher. Growing with them is paper birch, a successful "pioneer" tree where fire or wind have disturbed the forest. Mountain ash trees grow in sunny openings in the boreal forest and up into the alpine zone. Look for their white flowers in spring and orange berries in fall along the trail.

The boreal forest is home to fewer kinds of plants and animals than the hardwood zone. There are, however, more mosses and leafy liverworts here, intensifying the greenness.

A variety of birds are found in the boreal zone and all sing distinc-tive songs. "Trees, trees, murmuring trees," announces the beautiful black-throated green warbler. Boreal chickadees are brown-, not black-capped. Colorful golden-crowned kinglets call along the trail right up to the summits. If you are lucky, you may spot a white-winged or red crossbill, a black-backed woodpecker or even rarer three-toed woodpecker, both with yellow caps. Red squirrels scold you from their territories. Snowshoe hares hop by in summer brown or winter white. Recently hikers have sighted the scarce but increasing pine marten, a brown, tree-climbing member of the weasel family.

Balsam fir
Abies balsamea
The only Adirondack fir. Found up to 5300 feet on Mt. Marcy. "Bonsai" forms in alpine zone. Flat, "friendly" needles, blue-green color. Cones stand straight up. *Pine family*

Red spruce
Picea rubens
The major spruce of the Adirondack forest, found to 4000 feet. Needles sharp-pointed; cones hang down. Some trees reach over 200 years old in the Adirondacks. *Pine family*

Paper birch
Betula papyrifera
The major Adirondack pioneer tree, it comes in after fires and blow-downs. Northern hardwood and spruce-fir zones. Beautiful white, peeling bark. *Birch family*

Mountain ash ❀ JUN
Sorbus americana
This white-flowered tree has orange rather than the red-orange berries of close relative, *S. decora*. Trailside from spruce-fir to summit. Birds eat and distribute berries. *Rose family*

Bunchberry 🌸 JUN–JUL
Cornus canadensis
Looks like a miniature dogwood tree with 4–6 leaves in a whorl; 4 petal-like bracts and tiny central flowers that become bright red berries. In all Adirondack zones. *Dogwood family*

Snowberry 🌸 JUN
Gaultheria hispidula
Low trailside plant in the spruce-fir forest and krummholz. Alternate evergreen leaves, tiny bell-shaped flowers, and oversized berries eaten by hares and others. *Heath family*

Bluebead lily 🌼 MAY–JUN
Clintonia borealis
Northern hardwood and spruce-fir forest to timberline. Yellow flowers followed by blue bead-like fruits. In summer you can see both stages as you climb a peak. *Lily family*

Bluebead lily 🔵 JUL–AUG
Clintonia borealis
Also called Clintonia after New York governor DeWitt Clinton of Erie Canal fame. Oval-shaped leaves are smooth; similarly-shaped pink lady slipper leaves have pleats. *Lily family*

Twinflower ✿ JUN

Linnaea borealis

Evergreen, opposite, notched leaves; pink to white flowers. Truly circumboreal, including Lapland, where Linnaeus found it and named it after himself! *Honeysuckle family*

Starflower ❀ JUN–AUG

Trientalis borealis

Grows in all zones to alpine. Everything about this plant is starry, the whorl of 5–10 leaves and the 7 or more pointed white petals. "Borealis" means "of the north." *Primrose family*

Wood sorrel ❀ JUN–JUL

Oxalis montana

An indicator species of the spruce-fir zone, found from 1200 to over 4900 feet. Ground cover with pink-veined solitary white flowers, 3 leaflets. *Wood sorrel family*

Goldthread ❀ MAY–JUL

Coptis trifolia

Grows in all mountain zones; it blooms in alpine zone when in fruit lower down. Shiny evergreen leaves. Look in moss for its bright yellow (gold) "roots." *Buttercup family*

Rose twisted stalk ✿ MAY-JUN
Streptopus roseus
Found in northern hardwood forest
and higher. White-flowered twisted
stalk is characteristic of the spruce-
fir zone. Both have branched stems
and hanging flowers. *Lily family*

Painted trillium ❀ MAY-JUN
Trilliium undulatum
Grows in the northern hardwood
forest, but prefers spruce-fir forest.
The only white Adirondack trillium,
it has central magenta streaks on its
wavy petals. *Lily family*

Indian pipes ❀ JUL-AUG
Monotropa uniflora
Found in northern hardwood forest
to 4000 feet. A true flowering plant
without chlorophyll, it is a harmless
parasite, dependent on tree root
fungi for nutrients. *Indian pipe family*

Canada mayflower ❀ JUN
Maianthemum canadense
Found in colonies to timberline.
Blooming time depends on eleva-
tion. 6" high with fragrant white
flowers followed by spotted berries.
Lily family

SPRUCE-FIR FOREST

Large-leaved goldenrod ✿ AUG–SEP
Solidago macrophylla
Colorful trailside summer bloom
from spruce-fir forest to timberline,
also in alpine snowbanks. Up to 4
feet tall. Large leaves are noticeable
early in season. *Composite family*

Sharp-leaved aster ✿ AUG–OCT
Oclemena acuminata
Also aptly called "whorled wood
aster", it is a companion to the
large-leaved goldenrod—and to fall
hikers. Grows in spruce-fir forest
to timberline. *Composite family*

Oak fern
Gymnocarpium dryopteris
A characteristically Adirondack fern,
one of our loveliest. Small three-part
frond with a dark stem. Grows in the
northern hardwood and spruce-fir
forests up to timberline.

Three-lobed bazzania
Bazzania trilobata
Not a moss but a large leafy liverwort, a dominant ground cover in
the spruce-fir forest. Leaves in two
rows, with 3 teeth (shallow lobes).
One of many liverworts in this zone.

Steve Faccio

Red squirrel
Tamiasciurus hudsonius
Considers itself the rightful owner of the spruce-fir forest and scolds trespassers in its territory. This conifer-seed-eater is the most often seen and heard Adirondack mammal.

Porcupine
Erethizon dorsatum
Active at night, but seen in conifers along trails. Eats buds, twigs, and bark; also gnaws trail signs and lean-tos! Dens in hollow trees and rock cavities; does not hibernate.

Bill Silliker

Pine marten
Martes americana
A tree-climbing member of the weasel family; yellow-brown fir shading to dark brown on bushy tail. This predator has a long slender body (16–24").

Moose
Alces alces
Large, ungainly animal, the male with antlers up to 6 feet across. Once extirpated from Adirondacks, has now returned. Browses tree buds and bark, prefers aquatic vegetation.

SPRUCE-FIR FOREST

Periodically, sections of forest are destroyed by fire, windstorms, landslides, and other natural disturbances, and you can find evidence of these phenomena along the trail.

Fir waves are crescent-shaped bands of dead balsam firs, flanked on one side by declining mature forest, on the other by regenerating fir seedlings. Each band is actually a "wave" moving slowly through the forest with trees dying at the wave's leading edge. The speed of the wave movement, up to five feet per year, is highest on those ridge tops with the most prevailing wind exposure. Death, regeneration, and the maturation of a new fir forest is a natural cyclic process.

Hurricanes, avalanches, and fires have also altered the High Peaks landscape. The great 1950 blowdown buried many trails under "oceans of fallen tangled trees." Smooth glaciated rock and steep slopes make landslides a major Adirondack feature. One huge slide was observed on Colden in 1869. Gothics and Giant also show major slides and new ones continue to occur. Areas scarred by fire or windstorm are recolonized by stands of paper birch. Spruces and firs eventually grow in to replace them.

Bands of silvery dead trees mark leading edges of fir waves on Esther Mountain.

SPRUCE-FIR FOREST

Inside a fir wave mature trees die, fall, and are replaced by a new generation of young balsam firs.

Multiple slides on Mount Colden (left) and the aptly named Big Slide Mountain (above).

At treeline, firs and spruces are only a few feet tall.

KRUMMHOLZ

At about 4000 feet, the boreal forest composition begins to change. Mineral soils give way to organic soils and trees become shorter. Red spruce becomes scarce and you enter almost pure stands of balsam fir. Depending on their age and past disturbance, these fir forests may be thick, dark, and almost impenetrable. They can also be more open, with pink-striped wood sorrel flowers and big red-stem moss, *Pleurozium*, carpeting the forest floor. Other mosses and a leafy liverwort, *Bazzania*, are dominant here too. In areas previously disturbed by fire or blowdown you see young balsam firs growing up through stands of paper birch.

As you approach timberline, you enter a remarkable zone of moss, matted shrubs, and stunted, tangled trees—the krummholz or "crooked wood" zone. Red spruce cannot withstand the extreme environment here, but its close relative, black spruce, together with balsam fir, form the krummholz. This forest of dwarfed, often prostrate trees clings to the mountain at the limit of its endurance of short summers, thin soil, and winter winds.

Why these strange crooked trees, these flagged or broomsticked

Entering the krummholz: trees shrink down as you hike up.

Rime ice forms on windward side of trees, damaging branches.

Repeated damage by prevailing wind causes "flagging."

outposts of forest? Wind, together with abrasive wind-carried ice particles, are probably major factors. Exposed branch tips are repeatedly killed back, but protected side branches spread and grow into any gap. Wind flow is all-important. Trees in the same soil in a sheltered spot may be small, but well-shaped and without dead branches.

In the krummholz zone you see "flag trees" with branches killed on the side facing the prevailing wind. Others trees resemble broomsticks, with live branches at the top of a bare trunk. In these trees the crown has grown above the abrasion zone of windblown ice.

"Broomsticking" occurs when trees grow above the zone of abrasive windblown ice.

The cutoff line above which trees can no longer survive the harsh summit conditions is not usually an abrupt one. Depending on the slope, amount of soil, and prevailing winds, krummholz may persist to the mountain summit in protected patches among the true alpine, or tundra, vegetation. Black spruce, a tree of northern bogs elsewhere in the Adirondacks, forms snarled prostrate mats.

At timberline on Mt. Marcy the frost-free summer season is only two months long, compared to more than three months in nearby Lake Placid. In addition to low temperatures and the short growing season, fog further reduces photosynthetic rates. On the highest Adirondack peaks large areas of summit consist of exposed rock. Where soil does exist, it is likely to be too thin, peaty, or waterlogged for tree growth. Lack of snow cover, strong drying winds, and frost-heaving also make life impossible for trees, except for the true dwarf species like bearberry willow, only inches tall.

In this harsh environment, changes in the weather may be swift and severe. As you enter the alpine zone, be prepared for rain, cold, ice, or snow at any time of year.

Nancy Slack and alpine ecology students above treeline on Haystack.

KRUMMHOLZ

More than a dozen Adirondack peaks have alpine habitat, including Gothics in the Great Range.

THE ALPINE ZONE

Climb any one of sixteen Adirondack peaks, but most notably Marcy, Algonquin, Wright, Iroquois, Colden, Skylight, Haystack, Basin, Saddleback, Gothics, or Dix—or drive up Whiteface Mountain—and you will find some true arctic plants. These species blanket the treeless, arctic tundra landscape of northern Canada, Alaska, and Greenland. You have reached them here by climbing 2000–3000 feet upward into the alpine zone.

A combination of physical factors, geological history, and available post-glacial species have resulted in this very special landscape on

our highest Adirondack peaks. The alpine zone's climate, including mean annual temperatures, frost-free periods, exposure, and wind speeds, can be comparable to that of the arctic region. Frost-heaving damages roots and causes instability. Lack of sufficient soil and the presence of strong winds keep out not only upright trees, but many herbaceous and annual plants. Poor, peaty, and often waterlogged soils occur in both habitats.

There are differences as well. Summer temperatures may be higher in the Adirondacks, for example, and the day length at the summits is, of course, the same as at the trailheads, not continuous summer

Cushion-shaped diapensia is a true arctic plant.

37

Lying low—a prostrate bearberry willow hugs the rock for warmth and shelter.

daylight as within the Arctic Circle.

Many of the plants that grow here are rare or endangered. All of them are protected. Look carefully, take photos, but do not pick.

ALPINE PLANT COMMUNITIES

The alpine landscape may appear uniform at first, but observe carefully and you will discover a mosaic of different plant communities. You can see what looks like bare rock, but there are mosses and lichens on these surfaces and on the thin soil. You can see flat expanses, some that look like grassy meadows, some that are heath-like with dwarf shrubs. In depressions and out of the wind you can see pockets of dwarf trees.

Enjoy these alpine communities from the trails. The soils here quickly erode and lose scarce nutrients when exposed to foot traffic.

Tundra plants are well-adapted to invade a large variety of difficult microhabitats. Though the wind may be roaring at your ears, the microclimate near the ground surface is more conducive to plant life. Typically, it is warmer here; the soil, even if thin, absorbs solar radiation. The wind is reduced, especially inside a plant cushion. Small size and cushion-shaped forms are common plant adaptations to severe conditions.

Microclimate and topography often vary over a few feet of terrain, and plants group themselves into recognizable, though often intergrading, communities depending on the tolerances of each species. Here's a quick guide to plant communities to look for.

PIONEER COMMUNITIES

On exposed rock surfaces, both bedrock and boulders, lichens and mosses are the most successful pioneers, able to colonize bare rock. These plants can collect nutrients from rain and dust. Rock lichens come in many colors and contain acids that dissolve rock particles, fostering soil formation. Mosses can become established in rock crevices and spread, providing a seedbed for other plants. Some pioneer plants, like haircap mosses, alpine clubmoss, and mountain sandwort, can invade raw peat after a disturbance has removed the vegetation.

DIAPENSIA COMMUNITY

In the very most windswept sites, you will find one of our most beautiful alpine plants: diapensia, or mountain bride as it is sometimes called. If you brave the blackflies in early June you will see it covered with waxy white blooms. It often grows with other small shrubs such as bilberry, Lapland rosebay, and alpine azalea.

How can these plants survive this extreme environment? The rule is to stay low to the ground and tightly packed to avoid the wind. Diapensia forms a compact dark-green cushion which absorbs the sun and heat. Lapland rosebay, a miniature rhododendron, uses a

Mountain sandwort and mosses are plant pioneers in the alpine zone.

Algonquin meadows—Bigelow's sedge
and deer's hair sedge in late May (above)
and deer's hair sedge in August (right).

similar strategy. Note that there is
considerable open space in this
community; few plants can manage
these high winds, frequent frost
action, and lack of winter snow
cover. Sometimes the diapensia
cushions are eroded by the formation of needle ice, but diapensia
does reproduce and start new
colonies. See if you can spot a
seedling.

SEDGE MEADOW COMMUNITY

This is found on cold, north- and
west-facing slopes, gentle slopes, or
nearly level topography with high
soil moisture and a thin snow cover.

Although there are several true
alpine grasses, these meadows are
dominated by sedges, especially the
Bigelow's sedge (shown at top). It
is able to photosynthesize at low
light levels and can grow even in
very foggy sites. Look for it on the
summit of Whiteface Mountain.

Intertwined heath shrubs display their fabulous fall color.

Indian poke in an alpine snowbank community on Mount Marcy.

DWARF SHRUB/HEATH COMMUNITY

Where there is winter snow cover that melts early and the soil is well-drained, dwarf shrubs like blueberry, bilberry, black crowberry, and Labrador tea form a community. Dwarfed herbs such as bunchberry and goldthread, and the fruticose or shrubby brown Iceland lichen, often grow here too, in between the shrubs. The whole mat may be up to twelve inches deep, and in contrast to the diapensia community has almost 100 percent plant cover. There are no empty spaces; this community makes a compact interwoven mat.

SNOWBANK COMMUNITY

On the leeward sides of mountain peaks, out of the strongest winds and where the snow lies late, one can sometimes find another meadow community. Lingering snowbanks provide moisture and protection here. Look for Indian poke or false hellebore, a tall plant with pleated leaves. It seems out of place here, but grows with other snowbank plants such as hairgrass and alpine goldenrod, and many flowers like bluebead lilies and blue closed gentians which are usually found farther down the mountain.

Bog plants surround open water on the summit of Algonquin (above). Cotton sedge is one of several lowland bog plants found in the alpine zone (right).

ALPINE BOG COMMUNITY

Where there are poorly drained depressions with peaty organic soil, small alpine bogs develop. These are among the most interesting habitats. Look closely and you will see a small version of an Adirondack lowland bog, with many of the same plants in miniature form. There is sphagnum or peat moss, in brown, green, or raspberry pink, as a substrate for all the other plants. The pink-flowered small cranberry creeps along, accompanied by dwarfed forms of leatherleaf, bog laurel, Labrador tea, tiny carnivorous sundews, and a white-tufted sedge often called cottongrass. Sometimes the peat moss or sphagnum grows as much as half an inch a year, nearly blanketing the dwarf shrubs.

DWARF FOREST COMMUNITY

In sheltered depressions within the alpine zone one can find patches of dwarf trees. Black spruce, in its usually prostrate form, is able to propagate vegetatively. Very occasionally it bears cones. Balsam fir, usually more upright, has more frequent cone crops. Other dwarf trees such as larch, with its deciduous needles, and the alpine species of paper birch with its heart-shaped leaves, are also found in wind-protected depressions.

The Adirondack views are magnificent, when there is a view and you are not all fogged-in after a long climb. But there is much to be enjoyed by looking down as well as out. Over 100 different flowering plants, plus dozens of moss and lichen species, grow on Adirondack alpine summits. Some of these plants are rare, threatened, or truly endangered. These special plants are remarkably resilient to harsh weather, but, sadly, fragile underfoot.

A dwarf forest of balsam fir grows in a sheltered nook at 5000 feet.

Some alpine sites are too tough for flowering plants, but mosses and lichens may thrive. Two mosses, shag moss, *Racomitrium*, and granite moss, *Andreaea*, join lichens on rock faces. Haircap mosses form nearly soil-free communities. Sphagnum mosses form a substrate for the alpine bogs that develop in poorly drained depressions. Several rare or endangered mosses grow here and nowhere else in New York State. One, a pink, stringy peat moss, *Sphagnum pylaisii*, cascades over wet rocks. This species is one of very few plants that have increased on the summits in recent years, perhaps in response to acid precipitation. Recent experiments show that it grows well at the very acid pH of 3.

Lichens are complex organisms, the result of a mutualistic relationship between a fungus and an alga. They have important functions in the alpine ecosystem, particularly as soil formers. Some, which contain blue-green bacteria as well as green algae, can "fix" atmospheric nitrogen, converting it into a form that other plants can use.

Lichens can thrive in the toughest spots.

Lichens grow in a great variety of forms and colors; there are over 200 different species between the trailhead and summit of Algonquin. Some, like map lichen, are "crustose," forming a flat crust on rock surfaces. Others are "foliose" or leafy, and often grow on trees. Target lichen is a foliose rock lichen. Some are "fruticose" or shrub-like, such as the reindeer lichens and Iceland lichen which often grow in the ground layer of the dwarf shrub community. Look carefully and you may see the bright red fruiting bodies, or "redcoats," on the British soldiers lichens.

Peat mosses
Sphagnum
Sphagnum species in green, brown, pink, and red grow in the alpine zone, providing a home for miniature alpine bog species like sundew, cranberry, and cotton sedge.

Peat moss
Sphagnum capillifolium
Another beautiful peat moss, with small "pom-poms" of deep pink, in small hummocks in the alpine and krummholz zones. *S. fuscum* is a similar, but brown, alpine species.

Peat moss
Sphagnum pilaesii
This moss, rare elsewhere, forms beautiful pink clumps over moist rocks on the alpine summits of Marcy, Algonquin, and other high peaks. Also found in Lake Colden.

Granite moss *Andreaea rupestris*
Shag moss *Racomitrium heterosticum*
Both grow on rocks in the alpine zone and below. *Andreaea* is very drought-tolerant; it looks black and dead until revived by rain. *Racomitrium* is a brighter yellow-green.

Awned haircap moss
Polytrichum piliferum
Two hair cap mosses carpet the ground in the krummholz. This one has leaves with silvery hair points. *P. juniperinum*, juniper haircap moss, has red-brown leaf tips.

Jan-Peter Frahm

Brocade moss
Hypnum imponens
Beautiful creeping moss with neat pinnate stems "woven" together, the leaves all curled and turned to one side. At all elevations including sheltered niches in the alpine.

Big red-stem moss
Pleurozium schreberi
The dominant carpet species of the spruce-fir zone and in sheltered alpine sites. This showy, creeping feather-moss has an easily seen bright red stem.

Jan-Peter Frahm

Blue-green pogonatum
Pogonatum urnigerum
An attractive relative of the haircap mosses, common in the krummholz zone along the trail. Blue-green rosettes of opaque broad leaves; smooth, not angled, capsules.

Helmet moss
Conostomum tetragonum
A true alpine and arctic moss, found on 4 Adirondack summits. Blue-green with hair-pointed stiff leaves in 5 spirally twisted rows. Also in Greenland and the Canadian Arctic.

Beaked rock tripe
Umbilicaria proboscidea
This and related lichens are attached to rocks by a central holdfast on the underside. Several live in the alpine zone. Its Latin name refers to elephant skin—it has rough ridges.

Old man's beard lichen
Usnea sp.
These are the familiar grey lichens that hang from spruces and balsam firs. They do not harm trees—they are epiphytes, not parasites. There are several similar species.

Map lichen
Rhizocarpon geographicum
A conspicuous alpine rock lichen, forms yellow-green continent-like patches. This firmly attached crustose lichen is a very slow grower—only 5/8" per century!

Target lichen
Arctoparmelia centrifuga
Another conspicuous alpine rock lichen, not crustose but foliose, with leaf-like edges. Target-like because of concentric rings, the inner ones recolonized as the center decays.

Worm lichen
Thamnolia vermicularis
This hollow unbranched lichen lives in the Adirondack alpine zone, but is much more common in the Arctic. High Peak worm lichens are rare. Please don't disturb them!

Snow lichen
Flavocetraria nivalis
Light green with flattened lobes, rare in the Adirondack alpine zone. Found in late snowmelt areas. Common on Baffin Island and elsewhere in the Arctic where caribou eat it.

Gray reindeer lichen
Cladonia rangiferina
The common, upright, shrubby (fruticose) ground lichen that arctic reindeer and caribou eat. Grows in krummholz and alpine zones. The branchlets are swept to one side.

Star-tipped reindeer lichen
Cladonia stellaris
This attractive alpine lichen is the one used for trees in model railroads. Yellow-green rounded heads. *Cladonia mitis* and other relatives also live on the Adirondack summits.

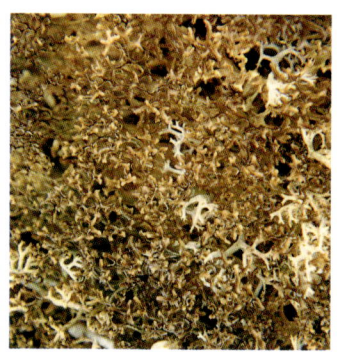

Iceland lichen
Cetraria laevigata
A common ground-dwelling fruticose lichen, important in the alpine shrub communities. Tan or brown, brittle with broad lobes. Look for it under bilberry bushes.

Foam lichens
Stereocaulon sp.
Stereocaulon species are commonly found on alpine rocks in the sun. Pale gray to white; low colonies with branched stalks covered with lobules or granules.

British soldiers lichen
Cladonia cristatella
Another fruticose lichen of the alpine and lower zones. Other alpine *Cladonias* also have red fruiting bodies (apothecia). Named for 18th-century British soldiers' red coats.

Red-tipped goblet lichen
Cladonia pleurota
This species has stalked cups with "soredia," asexual reproductive granules containing both fungi and algae. *C. coccifera* also has cups and larger red apothecia.

Look for these spore-bearing plants, many of them evergreen, in fall as well as summer. The mountain wood fern and the delicate oak fern grow right up to timberline and the northern beech fern is found on the summits in protected habitats. Bristly clubmoss runs along the spruce-fir forest floor and among the krummholz. Small clumps of alpine clubmoss are hardy pioneers where bare peat is exposed.

A great many species of ferns and fern allies, including both club-mosses and horsetails, are found in the Adirondacks. These plants reproduce by spores instead of flowers and seeds. They have a con-ducting system and can thus grow quite large. In the tropics there are even tree ferns! In past ages there were also giant clubmosses and horsetails, now found only as fossils.

Adirondack species include seven clubmosses, six horsetails, and thirty ferns. Few, however, ven-ture higher than 2500 feet. Among the ferns, the familiar polypody, Christmas fern, and sensitive fern of the northern hardwood forest drop out, but lady fern, interrupted fern, and intermediate wood fern, in addition to those pictured here, continue up the slopes.

Bristly clubmoss
Spinulum annotinum
Grows in spruce-fir forest to treeline;
at 4940 feet on Haystack! A trailing
clubmoss; spore-cones without
stalks. Similar staghorn clubmoss,
Lycopodium clavatum, has 12" stalks.

Alpine clubmoss
Huperzia appalachiana
Found in alpine and krummholz
zones; a pioneer species after
disturbance. Spore cases on special
stemleaves; asexual reproduction
by flap-like "gemmae." 4–12" high.

Narrow beech fern
Phegopteris connectilis
Found at all elevations into alpine
zone. Long triangular fronds with
lower leaflets pointing down; space
on stem between lowest leaflet pair.
Up to 12" high.

Mountain wood fern
Dryopteris campyloptera
Spruce-fir forest to alpine zone.
Frost sensitive, not evergreen.
Grows 12–24" high. Closely related
evergreen *D. intermedia* is found
in lower zones.

These grass-like plants have flowers, too, but since they are largely wind- rather than insect-pollinated, they do not have showy blooms. Some, like hairgrass, are true grasses; others, like the alpine bog-dwelling cotton "grass," are sedges. Grasses and rushes have round stems; those of most sedges have "edges"— angled stems. Many sedges and grass species grow on the Adirondack peaks. Some easy-to-identify alpines are shown here.

Alpine sweetgrass
Anthoxanthum monticolum
Attractive fragrant grass with 12" tufted, short leaves. Large florets with 1/4" awns. Look closely for these and enjoy the fragrance.

Hairgrass
Hierochloe alpina ssp. *orthanthum*
A common alpine grass that grows to 12" high. Also found at lower elevations. Hair-like leaves in dense tufts; florets have distinctive bent awns. Purplish or silvery sheen.

Highland rush
Juncus trifidus
Unlike grasses and sedges, rush flowers have petals and sepals— usually brown or green. Common in alpine turfs; 2–3 narrow leaves on top of 12" stem surround small spikes.

Bigelow's sedge
Carex bigelowii
This dominant sedge forms alpine meadows. Grows 12" high with purple spikes and yellow anthers in bloom. Jacob Bigelow was a famous early 19th-century botanist.

Deer's hair sedge
Trichophorum cespitosum
A dominant alpine sedge, forming large windswept patches. Green in summer, turning a tawny brown in fall, the color of a deer's hair. Grows in dense tufts, up to 12" high.

Mountain sedge
Carex scirpoidea
Found in alpine zone; an arctic species at its southern range limit in New York and New England. Unusual separate male and female spikes; grows to 12" high.

Cotton sedge
Eriophorum vaginatum, ssp. *spissum*
Also called "cotton grass." Found in both lower elevation and alpine bogs. Grows in tussocks with narrow leaves at base. This species has only one "head" (spike).

ALPINE ZONE

Few herbaceous or non-woody plants are restricted to the alpine zone, but many survive here, often in miniature form. Annual (one year) plants cannot grow here, but perennials with woody underground parts bear flowers and also fruit in the shorter summit growing season. Starflower, Canada mayflower, goldthread, wood sorrel, bluebead lily, bunchberry, and even Indian poke and bluets grow in sheltered places. Two late-bloomers, white wood aster and blue closed gentian, occur up to timberline. White-flowered three-toothed cinquefoil, characteristic of the alpine zone, is also found on ledges further down. Bright yellow alpine goldenrod, Boott's rattlesnakeroot, and mountain sandwort are strictly alpine flowers. Mountain sandwort forms small white-flowered cushions on exposed peat and gravel patches and blooms on nearly all the alpine summits from June to September.

Mountain sandwort ✤ JUL–SEP
Minuartia groenlandica
Grows in krummholz and alpine zones. Common along trails; a pioneer plant in areas disturbed by frost or boots. Forms tiny tufts and many-flowered clumps. *Pink family*

Three-toothed cinquefoil ✤ JUL–SEP
Sibbaldiopsis tridentata
Characteristic of the alpine zone, but also found in rocky sites on subalpine peaks. The three evergreen leaflets, each with three teeth, turn bright red in the fall. *Rose family*

Narrow-leaved closed gentian

Gentiana linearis ✿ JUL

A welcome late summer sight along subalpine trails. Grows 1–3 feet tall, prefers open areas. The flowers remain closed, but pollinating insects do get in. *Gentian family*

Indian poke ✿ JUN

Veratrum viride

Also called "false hellebore." A large plant (up to 2 feet tall) where snowbanks melt late. Pleated leaves and tall, branched flower stalks with many green flowers. *Lily family*

Rattlesnakeroot ✿ JUN

Prenanthes trifoliolata

This cut-leaved species is found at all mountain elevations; grows up to 4 feet tall. A dwarf alpine variety, also known as *P. nana,* has lobed leaves, unlike *P. bootii.* *Composite family*

Boott's rattlesnakeroot ✿ JUN

Prenanthes bootii

A rare dwarf alpine with unlobed leaves. Grows only on New York and New England summits. Named for Francis Boott, an early 19th-century botanist. *Composite family*

55

Round-leaved sundew ✿ JUN
Drosera rotundifolia
A tiny carnivorous plant found in peat moss of alpine bogs. Rounded leaf blades with sticky hairs. White flowers on taller stalks. Glands exude fluid to trap insects. *Sundew family*

Bluets ✿ MAY-JUN
Houstonia caerulea
Flowers in the alpine zone may be blue, occasionally white, with yellow centers. Tiny rounded opposite leaves, 2–6" high, in clumps. Prefers moist sites. *Madder family*

Harebell ✿ JUL-AUG
Campanula rotundifolia
Grows at all elevations to 4000 feet on cliffs and rock ledges. Nodding flowers, narrow stem leaves, and small rounded basal leaves. 6–20" high. *Bluebell family*

Alpine goldenrod ✿ JUL-AUG
Solidago leiocarpa (formerly *S. cutleri*)
This small goldenrod grows only in the alpine zone. Leaves are elliptic and toothed. Originally named for 19th-century botanical explorer Manassah Cutler. *Composite family*

Adirondack summits are tension zones where growing conditions become too severe for trees to thrive; however, dwarfed trees survive in protected spots. In addition to balsam fir, black spruce, and heart-leaved paper birch, you may find larch, a conifer whose tall counterparts are at home in Adirondack bogs. Look for mountain alder and bearberry willow, with large "pussy willow" catkins for its tiny size, along with two kinds of dwarf birch. White-flowered Bartram's shadbush and red-berried skunk currant may be found near timberline.

Dwarf shrubs are the most successful life form in the alpine zone. Many belong to the heath family. Two beautiful species, alpine azalea and Lapland rosebay, are true alpines; you can only discover them in the Adirondacks by climbing to the summits. Other shrubs, like lowbush blueberry, bog laurel, leatherleaf, and Labrador tea are also

Dwarf forms of cedar, juniper, and larch (shown below on Haystack) can be found at treeline.

found down below. The pink-flow-ered, trailing, small cranberry is equally at home in mountaintop or lowland bogs. Black crowberry, with its tiny leaves and black berries, also lives in the Arctic, as does June-flowering diapensia. Mid-June to early July is the best time to see the dwarf shrub flower display, but flowers are in bloom on the sum-mits even in October.

A dwarf mountain ash at 5000 feet along the Van Hoevenberg Trail (above). Tiny shrubs above treeline, like alpine azalea, boast beautiful flowers (below).

Black spruce
Picea mariana
A tree of swamps and the subarctic, it replaces red spruce at high elevations. Blue-green needles, hairy twigs. Like balsam fir, a "bonsai" tree in the alpine zone. *Pine family*

Mountain ash   ✿ JUN
Sorbus decora
Compound-leaved tree, found at upper elevations along trails. It is shade-intolerant and grows right up to the summits. White flowers, orange-red berries. *Rose family*

Heart-leaved paper birch
Betula cordifolia
Sometimes considered a variety of paper birch, the pioneer forest tree. This is shrub-like in the krummholz and alpine zones. Heart-shaped leaves, reddish bark. *Birch family*

Larch
Larix laricina
Also called "tamarack"; common in wet lowland sites, it turns brilliant yellow and loses its needles in fall. Shrub-like in the alpine zone, sometimes with small cones. *Pine family*

Bearberry willow ✿ JUN–JUL
Salix uva-ursi
A tiny creeping alpine tree with conspicuous large pink-flowering catkins. Small, smooth, toothed oval leaves, green above and paler beneath. *Willow family*

Bearberry willow 🍎 JUL–AUG
Salix uva-ursi
Large reddish fruits follow flowers on the catkins. Grows in large flattened mats on rocky summits; conspicuous gnarled woody "trunks." *Willow family*

Tundra dwarf birch ✿ JUN–JUL
Betula glandulosa
One of several alpine dwarf birches. Prostrate to two feet high. Scalloped, rounded leaves turn red in fall; large catkins. This and *B. minor* are endangered species in NY. *Birch family*

Peter Zika

Round-leaved alpine willow ✿ JUN
Salix herbacea
Tiny, creeping willow thought to be extirpated in the Adirondacks. The Algonquin population is gone, but a Summit Steward recently found it on another high peak! *Willow family*

Mountain alder ✿ JUN
Alnus viridis ssp. *crispa*
Dwarf shrub in the alpine zone with thick, doubly toothed leaves; with male catkins and cone-like persistent fruits. Buds not stalked (unlike lowland speckled alder). *Birch family*

Leatherleaf ❀ MAY–JUN
Chamaedaphne calyculata
This common lowland bog species is a miniature shrub in the alpine zone. Strings of bell-shaped white flowers; evergreen, rough, scaly, leathery leaves. *Heath family*

Labrador tea ❀ JUN
Rhododendron groenlandicum
A large shrub in lowland bogs, much smaller in alpine zone. Evergreen leaves with inrolled margins and woolly beneath. The tea tastes better in Labrador! *Heath family*

Bog laurel ✿ JUN
Kalmia polifolia
Another lowland bog plant turned small alpine shrub; also seen in krummholz zone. Bright pink flowers similar to those of mountain laurel; shiny evergreen leaves. *Heath family*

61

Bog bilberry JUN
Vacciniium uliginosum
A ubiquitous alpine shrub and a
close relative of the two alpine
blueberries. White-pink, bell-shaped
flowers. *Heath family*

Bog bilberry JUN
Vacciniium uliginosum
The light blue berries make good
animal food. The small leaves turn
lavender in fall. *Heath family*

Small cranberry JUN–JUL
Vaccinium oxycoccus
Cranberries on top of Algonquin and
Marcy? Yes, creeping in peat moss in
little alpine bogs. Evergreen in-rolled
leaves, bright pink flowers, and red
berries. *Heath family*

Lowbush blueberry JUN
Vaccinium angustifolium
Very widespread, the common lowland edible species. Narrow, pointed
leaves; blue fruit. *V. boreale,* dwarf
blueberry, is a tiny close relative only
6" high. *Heath family*

Skunk currant ✹ MAY–JUN
Ribes glandulosum
Shrub with palmately-lobed leaves
that smell like skunk when crushed.
White or pink flowers; red, bristly
fruit. The only high elevation currant, to 5000 feet. *Saxifrage family*

Meadowsweet ✿ JUL–SEP
Spiraea alba var. *latifolia*
Common lowland species; one of
the larger alpine shrubs. Appears to
lack adaptations for its severe environment. Conspicuous in flower;
coarsely toothed leaves. *Rose family*

Bartram's shadbush ✹ MAY–JUN
Amelanchier bartramiana
Our only subalpine shadbush; found
to 4500 feet. Only one to four
flowers per cluster. John and
William Bartram were 18th-century
botanical explorers. *Rose family*

Crowberry ✹ MAY–JUN
Empetrum nigrum
In spite of small, evergreen, heath-
like leaves, this belongs to a different
family from most alpine shrubs.
Mat-forming; also found on rocky
lower summits. *Crowberry family*

Diapensia   ❀ MAY–JUN

Diapensia lapponica

Climb to an alpine summit by June 15th to see these beautiful flowers. Compact dark evergreen cushions in the most windswept "arctic" sites. *Diapensia family*

Lapland rosebay   ✿ MAY–JUN

Rhododendron lapponicum

A spectacular June alpine bloomer. A true rhododendron, but a tiny shrub, under 6" high. Oval evergreen leaves with scurfy scales. Found in the Arctic, too. *Heath family*

Alpine azalea   ✿ MAY–JUN

Loiseleuria procumbens

Another dwarf alpine beauty with evergreen leaves. Abundant on Mt. Washington, NH. An Adirondack rarity found on only one of the 46 high peaks. *Heath family*

No matter where you are on an alpine summit, you are likely to see dark-eyed juncos flitting about. Their song is a bell-like trill often described as a musical sewing machine. Listen to a white-throated sparrow whistle its classic "Old Sam Peabody, Peabody." Blackpoll warblers sing "zee zee" and nest high up on the mountain. Yellow-rumped, magnolia, and many other colorful wood warblers breed in the alpine or the spruce-fir zone. Look and listen for ravens, noisy aerial acrobats that are becoming more common. Along the trail to timberline, the smallest bird with the longest song is the winter wren. It cocks its tail and sings its heart out. Walk quietly and you will hear the beautiful songs of the thrushes, especially Bicknell's thrush, near the summits.

The spruce grouse lives in the spruce-fir zone, not the alpine. It is mainly seen in Adirondack forested bogs, but also has been reported from Marcy Dam and on the lower slopes of Algonquin. It is currently under study; please report any sightings to the Department of Environmental Conservation (DEC).

Jeff Nadler

ALPINE ZONE

Dark-eyed junco
Junco hyemalis
Familiar mountaintop bird, all gray and white with pinkish bill. Flashes white tail feathers. Song a musical trill. Breeds in Canada and on peaks; winters in most of US. 6"

Jeff Nadler

White-throated sparrow
Zonotrichia albicollis
The bird most likely to greet you on the summits, with its whistled "Old Sam Peabody, Peabody." A Canadian and mountain breeder. Striped head and white throat. 7"

Jeff Nadler

Steve Faccio

Raven

Corvus corax

Coniferous forests and mountain cliffs. Much larger than a crow, 24" vs. 15", with longer bill and wedge-shaped tail. Distinctive deep croak. It is fun to watch its aerial acrobatics.

Black-backed woodpecker

Picoides articus

Two yellow-capped woodpeckers are seen in Adirondack subalpine forests. The three-toed woodpecker *(P. tridactylus)* has a barred, rather than all black, back. 9"

Warren Greene

Warren Greene

White-winged crossbill

Loxia leucoptera

The only red bird with white wing-bars and a crossed bill. Crossbills are able to extract seeds from conifers, including spruces high on peaks. Females are yellowish. 7"

Red crossbill

Loxia curvirostra

Darkish red (male) or greenish (female) bird with its distinctive crossed bill. Crossbills are irregular visitors and breeders in the Adirondacks. 6"

Spruce grouse
Falcipennis canadensis
A rare dusky speckled grouse with chestnut tail tip. Feeds on evergreen needles and buds in lower spruce-fir forests; longer-tailed ruffed grouse is found in hardwood forests. 13"

Boreal chickadee
Poecile hudsonica
In the spruce-fir zone and higher you may find a chickadee with a brown cap and white cheek patch, unlike the black-capped chickadee. Trilled song. 5"

Golden-crowned kinglet
Regulus satrapa
Tiny bird, with striped face and yellow cap found up to treeline, higher than its ruby-crowned cousin. An acrobat, its song is a series of thin, high notes and a chatter. 4"

Bicknell's thrush
Catharus bicknelli
Breeds on Adirondack summits, unlike gray-cheeked thrush which looks similar. A rare species, currently under study in the Adirondacks. Beautiful distinctive song. 6"

Jeff Nadler

Jeff Nadler

Magnolia warbler
Dendroica magnolia
Many colorful wood warblers breed
in the spruce-fir zone; this beauty
often goes higher. Black "necklace"
and stripes on bright yellow. Sings
"sweeter, sweeter, SWEETest." 5"

Yellow-rumped warbler
Dendroica coronata
Common summer resident from
spruce-fir zone to treeline. Bright
yellow rump patch. Song a soft
inconclusive warble. Winters farther
north than other warblers. 5"

Jeff Nadler

Eric Dresser

Blackpoll warbler
Dendroica striata
The warbler most likely to be seen
high up, from 3000 feet to alpine
zone. Nests in spruce branches.
Male has distinct black cap. Song:
high notes on a single pitch. 5"

Winter wren
Troglodytes troglodytes
A tiny bird with a very long, complex, trilled song, that greets you
along trails up to the alpine zone.
Hunts insects on the spruce-fir forest
floor. Tail raised vertically. 4"

MAMMALS, AMPHIBIANS, & INSECTS IN THE ALPINE ZONE

Raccoons and their young are often seen near the trails, especially if you are still hiking at dusk. Porcupines, common at lower elevations, occur up to timberline. They eat bark, and you can find evidence of their chewing on krummholz trees. Snowshoe or varying hares, white in winter, brown in summer, are found all the way to the summit. They munch on tundra plants and leave their round pellet droppings behind. Rare long-tailed shrews also inhabit the alpine zone. With luck you may spot a marten; once very rare, they are now seen on the high slopes of Marcy and Algonquin.

A number of amphibians, such as American toads, red efts, two-lined salamanders, spring peepers, and wood frogs, can be found at high elevations, some of these up into the alpine zone. No reptiles are found above treeline.

Monarch and tiger swallowtail butterflies flit over the summits. On a sunny day look for different species of bees (four wings) and flies (two wings) busy pollinating the alpine flowers.

Eric Dresser

Snowshoe hare
Lepus americanus
A common Adirondack mammal, even above treeline. The name comes from its large feet. Brown in summer, turns white in the snow season. It eats twigs, buds, and bark.

Olivier Gilg

Short-tailed weasel
Mustela erminea
This small predator is 8–12" long. Called "ermine" while in its white winter coat with black-tipped tail. A carnivore, it eats mostly rodents. Dark scat often seen on trail rocks.

Mourning cloak
Nymphalis antiopa
Deep purplish brown, yellow border; look for blue spots near border. This is the first butterfly one sees at all elevations in spring. It overwinters as an adult.

Tiger swallowtail
Papilio glaucus
This unmistakable butterfly is the one you are most likely to see in the alpine zone. Look and discover which flowers it pollinates. Found from Newfoundland to Florida.

Insect pollinators
Flies, bees, wasps, and butterflies are surprisingly common in the alpine. They pollinate many flowers, including these inner true flowers of the bunchberry.

Common water strider
Gerris remigis
Although it cannot fly, this common lowland insect is found in pools and puddles above treeline up to 5000 feet. It preys on living and dead insects on the water surface.

CONSERVATION

Alpine plants are hardy; they cope with the most extreme of environmental conditions. But they do not cope well with hikers' boots. Their plant communities are fragile. If you step or sit on them you wear away the plant mat, exposing the thin soil to erosion by wind and water. There are at least twenty rare, threatened, and endangered flowering plant species on the Adirondack summits and many equally rare mosses and lichens. The round-leaved alpine willow was extirpated from Algonquin Peak. A new population was recently found by a Summit Steward on another high peak, but its future is uncertain. Several other species have been extirpated from their summit "islands" in the past fifty years.

To preserve these very special communities and the plants that occupy this habitat, the Summit Stewardship Program was developed. This cooperative program,

Summit Stewards explain ecology and conservation to hikers on alpine peaks.

Rock paving protects fragile alpine soil. Informed hikers stick to the trail. Rocks carried to the summits by hikers are used to build cairns and stabilize erosion.

jointly sponsored by the Adirondack Chapter of The Nature Conservancy (TNC), the Adirondack Mountain Club (ADK), and the New York State Department of Conservation (DEC), places uniformed naturalists on the summits. The Summit Stewards greet you, rain or shine, on peaks threatened by recreational misuse and explain the significance and fragility of the alpine communities. They encourage all hikers to stay on trails and on rock surfaces and thus off the plants to prevent further erosion and extinction.

Ed Ketchledge and other forest ecologists, along with hundreds of volunteers, pioneered a successful restoration program for eroded alpine summit vegetation. Rocks were first placed on damaged trailsides. Then grasses not native to the summits were planted in eroded areas, thus stabilizing the site. Gradually mosses invaded and provided a seedbed for mountain sandwort and other native alpine plants. The introduced grasses, which require fertilizer in these nutrient-poor soils, died out. This program was continued for over twenty years. It succeeded, but now all of us who hike to the alpine summits must become caring stewards of this unique and beautiful environment.

New techniques are underway to preserve the fragile alpine habitats. The Summit Stewards are doing a remarkable job in all kinds of weather to see that the summit rules are followed and to answer hikers' questions. They are also keeping track of the phenology of the alpine plants; i.e., what flowers bloom when, and when they are in fruit (see page 76). Many people from the New York State Natural Heritage Program, the Adirondack Nature Conservancy, and DEC help with their training at the start of each season.

Rocks have been important in recent projects. Summit Stewards are using them in a variety of ways to protect alpine habitats. These

HOW YOU CAN HELP

▷ Stay on the trails. Follow markers, painted blazes, and cairns—the rock stacks that mark the trails.

▷ Walk on solid rock whenever possible. Avoid stepping on vegetation, exposed soil, and gravel.

▷ Do not pick or remove any plants.

▷ Do not use vegetation to assist in climbing.

▷ Do not camp above 4000 feet, even in winter. Over 3500 feet, camp only at campsites designated by DEC.

▷ Avoid hiking above 4000 feet during spring thaw when alpine soils are most vulnerable. In addition, DEC may ask hikers to avoid hiking above 3000 feet during this season. Please honor these voluntary closures.

▷ Learn more about alpine ecology and Adirondack natural history. Talk with a Summit Steward, ask questions.

▷ Do not litter — if you carry it in, pack it out.

▷ Dogs must be leashed and under your immediate control at all times. In the alpine zone, keep dogs on the rocks and off the sensitive vegetation and exposed soil.

▷ Share these suggestions with others.

Rocks carried by hikers are used to build cairns and stabilize erosion.

rocks need to be carried up from lower elevations and many hikers are helping—perhaps some of you who read this book. A sign recently seen on Algonquin Peak reads:

Please leave your rocks here. Thank you for helping to protect the alpine vegetation! Those of you that stuck it out and helped carry rocks all the way to the top should be proud. Summit Stewards will use your rocks to define trails across the summit, build sturdy cairns and stabilize fragile alpine soils. If you opted not to bring a rock to the summit, you can help Summit Stewards and alpine plants by walking only on the solid rock surfaces. TNC and ADK

PHENOLOGY

The term phenology usually refers to the responses organisms make to recurring seasonal changes in their environments. In seasonal cycles, many environmental factors change: light intensity, snow cover, the relative lengths of day and night. Day length is the same on any particular calendar day every year, thus it is a dependable cue for both plants and animals. In the alpine zone, seasonal changes are more extreme than elsewhere: higher winds, rime ice, great temperature shifts, and dramatic soil movement caused by repeated freezing and thawing. All of these factors affect alpine plants.

As you climb the High Peaks you can observe these altitude and seasonal changes. In July, Clintonia may have blue berries at 2000 feet, yellow flowers as you continue up, and perhaps only buds in the alpine snowbank communities. At the trailhead you may find a bunchberry with ripening red fruit. At treeline it will still have its showy white bracts and flowers.

Be observant of this phenology. Report your observations of alpine plants to a Summit Steward, who will be keeping records of the many rare species.

Elevation matters—bunchberry may bloom in early June at 1500 feet, but not until mid-July at treeline when they are in fruit down below. Observe the stages of different species as you climb. Flowers that have gone by at the trailhead may still be blooming up high. Both these photographs were taken on the Van Hoevenberg Trail on the same day in late July.

Use this chart as a reference aid to quickly identify an unknown alpine zone plant in flower. The chart lists many of the species described in this guide.

PG.	SPECIES	MAY	JUN	JUL	AUG	SEP
25	Bluebead lily	▬	▬			
26	Starflower	▬	▬			
26	Goldthread	▬	▬			
56	Bluet	▬	▬			
61	Leatherleaf	▬	▬			
63	Skunk currant	▬	▬			
63	Bartram's shadbush	▬	▬			
63	Crowberry	▬	▬			
64	Lapland rosebay	▬	▬			
64	Alpine azalea	▬	▬			
64	Diapensia	▬	▬			
27	Canada mayflower		▬			
55	Boott's rattlesnakeroot		▬			
55	Rattlesnakeroot		▬			
56	Round-leaved sundew		▬			
59	Mountain ash		▬			
61	Bog laurel		▬			
61	Labrador tea		▬			
62	Bog bilberry		▬			
62	Lowbush blueberry		▬			
25	Bunchberry		▬	▬		
60	Bearberry willow		▬	▬		
62	Small cranberry		▬	▬		
55	Indian poke			▬		
56	Harebell			▬	▬	
54	Three-toothed cinquefoil			▬	▬	
63	Meadowsweet			▬	▬	
54	Mountain sandwort			▬	▬	
28	Large-leaved goldenrod				▬	▬
56	Alpine goldenrod				▬	▬

INDEX TO SPECIES

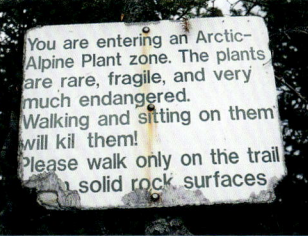

You are entering an Arctic-Alpine Plant zone. The plants are rare, fragile, and very much endangered. Walking and sitting on them will kill them! Please walk only on the trail solid rock surfaces.

Summit of Mount Marcy, 2006, with restoration area in right foreground where soil has been worn away by hikers' feet. Compare to photograph on page 14.

SELECTED REFERENCES

Kudish, Michael. 1992. *Adirondack Upland Flora: An Ecological Perspective.* Chauncy Press. Saranac Lake, NY.

McMartin, Barbara. 1994. *The Great Forest of the Adirondacks.* North Country Books, Utica, NY.

Mitchell, Richard S. and Gordon C. Tucker. 1997. *Revised Checklist of New York State Plants.* New York State Museum: Albany, NY. Updates for New York plant names can be found at http://atlas.nyflora.org.

Newcomb, Lawrence. 1977. *Newcomb's Wildflower Guide.* Little, Brown and Co.: Boston, MA.

Owen Sound Field Naturalist. 2002. *A Guide to the Ferns of Grey and Bruce Counties, Ontario.* Stan Brown Printers: Owen Sound, Ontario.

Peterson, J.M.C., and Gary Lee. 2004. *Birds of Hamilton County, NY.* Lake Pleasant, NY. (Also *Birds of Clinton Co.* 2nd edition by Mitchell and Krueger and *Birds of Essex Co.* 3rd edition, 1999, by Carleton and Peterson.)

Pope, Ralph. 2005. *A Hiker's Guide to Treeline Zone Lichens of the Northeastern States.* University Press of New England: Lebanon, NH.

Schottman, Ruth. 1998. *Trailside Notes: A Naturalist's Companion to Adirondack Plants.* Adirondack Mountain Club: Lake George, NY.

Slack, Nancy G., and Allison W. Bell. 2006. *AMC Field Guide to the New England Alpine Summits.* Appalachian Mountain Club: Boston, MA.

Steele, Frederic. 1982. *A Nature Guide to the Mountains of the Northeast.* Appalachian Mountain Club: Boston, MA.

Waterman, Laura, and Guy Waterman. 1982. *Forest and Crag: A History of Hiking, Trail Blazing and Adventure in the Northeast Mountains.* Appalachian Mountain Club: Boston, MA.